Revolutionizing Manufacturing

The Application of AI in Industry 4.0

Table of Contents

Chapter 1. Introduction

Welcome to this Special Report on "Revolutionizing Manufacturing: The Application of AI in Industry 4.0". Through our clear and insightful examination, we embark on a journey to the very heart of the manufacturing industry's transformation. Don't fret, we've made sure to demystify all jargon and reveal the nuggets of practical wisdom hidden beneath the technical veneer. Uncover how Artificial Intelligence-powered systems are not just the future, but the present reality, dramatically reshaping production lines worldwide. Whether you're an industry professional, an academic, or a curious observer, this report will equip you with a deep understanding of an unfolding industrial revolution—all while breaking down complex topics down into digestible chunks of everyday language. Get ready to uncover the momentum behind Industry 4.0, where the futuristic now meets the practical. By the end of this special report, you'll have unprecedented insight into this bold new frontier of manufacturing, making you a well-informed participant in these exciting times of technological advancement.

Chapter 2. Setting the Stage: The Manufacturing Landscape Pre-Industry 4.0

The manufacturing landscape has come a long way since the first industrial revolution. The keystone of this evolutionary journey is traced back to the mid-18th century—the days of steam and water power. From humble beginnings, technology has constantly shaped production lines, paving the way for unprecedented levels of industrial efficiency.

2.1. Chronology of Industrial Revolutions

From steam power to electricity, assembly lines to automation, industry has undergone various iterations of industrial revolutions. Each subsequent revolution building on its predecessor, but also radically transforming industrial practice, economics, and society as a whole.

1. The first industrial revolution (Industry 1.0) in the late 18th century replaced manual labor with mechanized production through water and steam power. This era marked the beginning of factory system.

2. Industry 2.0 started in the late 19th century, with the advent of electricity, enabling mass production capabilities via the division of labor and assembly lines.

3. The third revolution, taking shape in the late 20th century—Industry 3.0—was characterized by the integration of computers and automation into the production processes.

Before we can elucidate the features of Industry 4.0, we must comprehend the characteristics of the manufacturing landscape just prior its advent, essentially focusing on the late stages of Industry 3.0.

2.2. Understanding Industry 3.0

The term 'Industry 3.0' bears associated with the digital revolution. It represented a paradigm shift in manufacturing landscape, driven by developments in electronics, information technology, and automation. A simple assembly line was no longer enough—computers and automated processes began to dominate, rendering manual control obsolete. This shift fostered greater precision, operational efficiency, and high-volume production capability.

Some of the distinguishing technologies of Industry 3.0 include Programmable Logic Controllers (PLCs), Computer Numerical Control (CNC) machines, automation, and robotics. All of these innovations enhanced the precision, pace, and output of manufacturing processes.

Further fueling this revolution, computers did not just facilitate automation–they also contributed to advanced analytics and data processing. This provided decision-makers with deeper, data-driven insights, fostering informed strategic planning with less room for error.

2.3. Implications of Industry 3.0 on Manufacturing

The digital revolution irrevocably altered the manufacturing landscape, yielding myriad implications on various fronts.

1. Enhanced Productivity and Efficiency: With automated systems and robotics shouldering a significant workload, operations

became more seamless, swift, and efficient. Human error—a bottleneck in the production line—was drastically reduced, contributing to higher quality standards and better productivity.

2. Flattened Organizational Structures: Technology integration into manufacturing processes eliminated the need for hierarchical structures, favoring horizontal ones instead. Employees, albeit fewer in number, were now specialists handling complex machinery.

3. Globalization: The marvel of networked communication empowered businesses, facilitating global trade and collaboration. It dissolved geographical barriers, enabling companies to have global reach.

Despite the momentous strides of Industry 3.0, challenges remained. Manufacturing plants, while automated, were largely disconnected from one another. Interoperability was sparse, and system-wide resilience to shocks—such as a faulty production line—was relatively weak. Limited contextual awareness and decision-making ability of systems persisted as an impediment to further advancements in efficiency, flexibility, and productivity, which was halted by the predetermined instruction sets of machines—they worked in yes/no logic and lacked the capability to make intelligent decisions.

In essence, the factories of the Industry 3.0 era, while exponentially more innovative than their predecessors, required significant human intervention and oversight.

This transition phase, teetering between the established digital order of Industry 3.0 and the beckoning intelligence era of Industry 4.0, was laden with technological promise. A promise poised to usher an era of self-managing, connected, and intelligent manufacturing plants, thus setting the stage for Industry 4.0.

The advent of Industry 4.0 hence represents not just the next step on the ladder, but a quantum leap into an intelligent self-managing system that embarks on a new era of manufacturing. It is where the lines blur between digital, physical, and biological spheres. Harnessing the potential of technologies like Artificial Intelligence (AI), Machine Learning (ML), Internet of Things (IoT), and more, Industry 4.0 is the synthesis of cyber-physical systems that revolutionizes the manufacturing landscape in unprecedented ways.

As we recognize the transformative years of Industry 3.0 and its impact on present day manufacturing, we equip ourselves to better understand, appreciate and navigate the revolutionary tide of Industry 4.0. It's here where we observe the winds of change whispering the onset of a new industrial age—an age where machines breathe intelligence and industries pulse with connectivity. As we stand on the precipice of this new age, we also stand poised with the power to sculpt its course...a task underscoring the importance of situating our understanding of Industry 4.0 in the context of its industrial evolution.

Chapter 3. Demystifying the Concept: What is Industry 4.0?

The transition from one industrial revolution to another has always been a significant milestone in the history of mankind. It was quite a leap to move from the steam-powered machinery of the first industrial era to Industry 2.0's assembly lines, and subsequently to the automation and computerization of Industry 3.0. Yet, here we are today standing at the precipice of yet another revolution, one that is transforming the manufacturing sector unlike anything before – Industry 4.0.

3.1. Defining Industry 4.0

To begin, let's dive into the definition of Industry 4.0. It is the current trend of automation and data exchange in manufacturing technologies. Predicated on the pillars of interconnectivity, automation, machine learning, and real-time data, Industry 4.0 marries physical production and operations with smart digital technology, machine learning, and big data to create a more holistic and better-connected ecosystem for companies that focus on manufacturing and supply chain management.

3.2. The Pillars of Industry 4.0

The tenets of Industry 4.0 center around nine key technological components.

1. **Cyber-Physical Systems (CPS):** These bridge the gap between virtual and physical components allowing for interaction and cooperation between machines, humans, and smart systems.

2. **IoT (Internet of Things)**: The vast network of physical devices communicating and sharing data, providing real-time analytics and enabling faster and more accurate decisions.

3. **Cloud Computing**: Centralized data storage and processing power available on demand. This helps in enhanced data analytics, real-time processing, and accessibility.

4. **Cognitive Computing**: Advanced systems designed to mimic the human brain's process of acquiring knowledge to solve complex problems.

5. **Artificial Intelligence**: Helps in decision-making and predictive analysis, drives automation and enhances business efficiency.

6. **Augmented Reality**: This technology can help in process visualization, maintenance, and training in the manufacturing space.

7. **Additive Manufacturing**: More commonly known as 3D printing, this process builds products layer by layer, increasing flexibility in production.

8. **Big Data Analytics**: In this context, it is the application of advanced analytic techniques against large and diverse data sets from various sources.

9. **Autonomous Robots**: Robots capable of working collaboratively with human beings, and also learning from humans to operate more efficiently.

3.3. The Evolution towards Industry 4.0

The journey to Industry 4.0 has been evolutionary, not revolutionary—it's built upon advancements made during the previous industrial revolutions. The First Industrial Revolution saw the transition from skilled artisans making handmade items to mechanical production using water and steam power. The Second ushered in the era of mass production and assembly lines powered by electricity. The Third brought us the digital revolution with the advent of computers and automation. Today, the Fourth Industrial revolution, or Industry 4.0, builds on the digital revolution and combines it with physical systems to create integrated systems.

3.4. The Impact of Industry 4.0

The implications of Industry 4.0 are far-reaching. Firstly, it empowers businesses with real-time data and intelligence. The ability to gather, analyze, and act upon data in real-time provides unprecedented efficiency, productivity, agility, and flexibility.

Secondly, it allows the creation of what is known as a 'smart factory'. These factories, aided with technology, are able to run themselves with a reduced need for human intervention, giving rise to the possibility of lights-out manufacturing.

Thirdly, there is product customization. Industry 4.0 makes it feasible for manufacturers to produce highly personalized products based on individual customer specifications without sacrificing production efficiency.

3.5. Industry 4.0: The Future is Now

While it may sound like the stuff of science fiction, all aspects of Industry 4.0 are in practice today, albeit at varying stages of maturity. There is a compelling case to be made for the transformative power of these living-breathing cyber-physical systems in reshaping the economy, workforce, and society itself.

From this broad exploration of Industry 4.0, it's clear we are in the midst of a profound shift. The future of manufacturing lies in these integrated smart systems powering everything from supply chain logistics to sales and marketing initiatives to customer engagement.

It's a thrilling time to be in manufacturing and to participate in the world's next big industrial revolution, which, incidentally, is happening right now.

The next question then is, if Industry 4.0 is the present and future of manufacturing, how does Artificial Intelligence factor into this equation? That's precisely what the next chapter will delve into: the role of AI in this exciting new era for industry. So, buckle up for the next phase of our exploration.

In conclusion, Industry 4.0 represents an important shift in the manufacturing sector—embracing the potential that comes with intelligent systems, not just for efficiency and profitability, but for real transformation. The dawn of this evolution carries with it the promise of extraordinary opportunities, available to those willing to embrace the journey of change, adaptation, and innovation. With such insights, one can assert firmly that Understanding Industry 4.0 is critical to prospering in the future of manufacturing and business at large.

Chapter 4. A Deep Dive Into Artificial Intelligence

Artificial intelligence (AI) represents the epitome of modern technologic progression. Its roots stem back to the mid-20th century, with the dawn of computers, but now, we've entered an era where AI is profoundly transforming the way we produce, deliver, and consume products and services globally. From self-driving cars to voice-enabled personal assistants, AI applications are pervasive—virtually omnipresent in our daily lives. Whether you realize it or not, they're instrumental in several sectors, with manufacturing being no exception.

4.1. What is Artificial Intelligence?

Artificial intelligence refers to the simulation of human intelligence processes by machines, especially computer systems. These processes include learning, reasoning, problem-solving, perception, and language understanding. If we unbox this definition, the first prong to emerge is 'Machine Learning' (ML), an AI subfield focused primarily on the design of systems, enabling them to learn and make predictions based on some past experience.

Machine Learning uses algorithms to find patterns in data without the requirement of explicit programming. They adapt by themselves through experience (just like people). The more the algorithm is used, the better it gets at predictions. Within ML, there is another subset - 'Deep Learning'. Picture a set of matryoshka dolls; AI is the largest one, with ML concealed inside, and deep learning nestled even further in. Deep learning is a neural network with three or more layers. These neural networks attempt to simulate the behavior of the human brain—albeit far from matching its ability— to 'learn' from large amounts of data.

While a neural network with a single layer can still make approximate predictions, additional hidden layers can help optimize accuracy. When you hear the term 'deep' in 'deep learning', it symbolizes these layers. They serve to construct a good prediction model using the neural network, which is strung together with a vast number of nodes akin to neurons.

4.2. AI: The Heartbeat of Industry 4.0 in Manufacturing

AI's role in the fourth industrial revolution, Industry 4.0, is significant. This new phase in industrial evolution relies heavily on interconnectivity, automation, machine learning, and real-time data. If Industry 4.0 were a living organism, AI would, without doubt, be its central nervous system.

Manufacturing industries have leveraged AI in several practical applications. Predictive maintenance, enabled by Machine Learning, allows companies to predict when or which equipment may fail. It saves both time and money that would otherwise be lost due to unexpected equipment failure. Robotic Process Automation (RPA), another significant application area, automates repetitive and mundane tasks, thus freeing up human labor for more complex problem-solving activities.

Supply chain optimization also demonstrates another space where AI plays a noteworthy role. By analyzing vast volumes of data, AI can deal with demand and supply fluctuations, manage supply chain risks, and assist with strategic decision making.

4.3. The Evolution of AI: A Historical Perspective

The concept of artificial intelligence is not a new construct of the 21st century. Historically, its roots go back to the ancient world. Philosophers tried to explain human thinking as a symbolic system, an idea which formed the hypothetical underpinnings of artificial intelligence.

The actual term "Artificial Intelligence" was first coined by John McCarthy at the Dartmouth Conference in 1956. It was a significant event where top minds concerned with neural networks and automata theory congregated to discuss the possibility of building machines capable of simulating human intelligence.

Thereafter, the world of AI traveled through periods often referred to as "AI winters," where progress seemed to slow down, mainly due to reduced funding. However, the perception of AI changed dramatically in the 1990s and early 2000s, with the emergence of systems capable of beating humans at chess and quiz shows, flaring up the public's wonderment and curiosity about AI all over again.

Today, we stand at the verge of yet another sea change in the journey of AI. Leveraging unprecedented leaps in data volume, computational power, and advanced algorithms, we are creating AI systems smarter and more capable than ever before.

4.4. Challenges and Ethical Considerations

While discussing AI and touting its benefits, it's only fair to shed light on some of the challenges and ethical considerations it brings along. AI's adoption isn't without issues.

First, there's the challenge of the skills gap. Not everyone understands AI. It isn't just about having technical skills but about understanding AI's capabilities and how to leverage them strategically in business.

Data privacy is another significant concern. As AI systems depend on gargantuan amounts of data, ensuring privacy becomes more challenging. Companies need to abide by legal norms strictly because a breach could lead to a loss of consumer trust and potential legal implications.

Then comes the ethical aspect. AI systems do not inherently understand ethical boundaries. Training AI systems to differentiate right from wrong – now and in different cultural, social, or normative contexts – remains an ongoing quest in shaping a safe AI future.

4.5. Conclusion

Artificial Intelligence has gingerly advanced to a position of being a cornerstone in the manufacturing industry but equally in every aspect of our lives. Ranging from autonomic cars, market prediction models, to sophisticated manufacturing processes, AI has been a catalyst to exceptional development. At the same time, it invites us to ponder and address its latent challenges and ethical considerations that tag along. In all its grandeur, AI stands as a testament to the pinnacle of human technological achievement – a tool that we've shaped and now helps us shape the world around us in ways that were once only fodder for science fiction. To condense it all into a single thought - we stand at the precipice of a future where AI is no more a tool but an extension of our collective intellect.

Chapter 5. The Intersection of AI and Industry 4.0

Nestled at the core of Industry 4.0, the fourth industrial revolution, we find an incredibly powerful force—Artificial Intelligence (AI). This groundbreaking technology is not just shaping the future; it is actively redrafting the present. Through its vast potential and unparalleled capabilities, AI is surging to the forefront of industrial transformation, deftly merging the digital and physical realms in ways we've only dreamed of.

5.1. The Definition of AI and Industry 4.0

To begin, let's firstly understand the pillars we're dealing with. Artificial Intelligence, or AI, refers to the simulation of human intelligence processes by machines, especially computer systems. These processes include learning, reasoning, problem-solving, perception, and language understanding. The output is essentially a machine with the capability to handle tasks that would typically require human intellect.

On the other hand, Industry 4.0, also known as the fourth industrial revolution, signifies the current trend of automation and data exchange in manufacturing technologies. It encompasses cyber-physical systems, the Internet of things (IoT), cloud computing, and cognitive computing—all epitomizing the idea of smart factories.

5.2. Why AI is Integral to Industry 4.0

Industry 4.0 is all about automation and interconnectivity; it thrives on insights gleaned from machine data to enhance manufacturing processes. AI lies at the heart of this, enhancing the capacity to manipulate and analyse complex sets of information. Whether it's streamlining workflows, predicting maintenance, or optimizing energy consumption, AI enables smart factories to function efficiently and sustainably, react quickly to demands, and deliver superior quality products.

The combined force of AI and IoT leads to IIoT (Industrial Internet of Things), a leading driver of Industry 4.0. AI-powered algorithms can address massive IIoT-generated datasets, leading to actionable insights for making informed decisions. They complement each other, with IoT providing the data and AI offering the analysis power.

5.3. Practical Applications of AI in Industry 4.0

At this juncture, we can identify several practical applications wherein AI significantly boosts Industry 4.0 processes:

5.3.1. Predictive Maintenance

Predictive maintenance involves using machine learning algorithms to monitor equipment status and performance continuously. This enables factories to predict potential failures or malfunctions, allowing for timely maintenance activities and prevention of costly downtimes. It dramatically enhances efficiency and saves costs in the long run.

5.3.2. Quality Control

AI can exponentially enhance quality control measures. Advanced machine learning algorithms analyze product dimensions, raw materials, operating conditions, and other factors to detect anomalies in products. Over time, the system learns to identify faults more accurately, leading to substantial improvements in product quality.

5.3.3. Supply Chain Optimization

AI can analyze past data, current trends, and predictive analytics to optimize supply chain management. It can help determine the best transport routes, manage inventory efficiently, predict demand, and identify potential disruptions through pattern recognition, saving valuable time and resources.

5.3.4. Real-Time Decision Making

AI's ability to analyze an enormous amount of data in a short period allows for real-time decision-making. This feature is particularly useful in processes such as production planning, where immediate adjustments can be made based on current demand, supply chain status, material availability, and machinery conditions.

5.4. Overcoming Challenges with AI in Industry 4.0

While weaving AI into Industry 4.0 can gift us a host of benefits, it's not without its set of challenges. Data security, lack of skilled personnel, and resistance to change are all common obstacles that businesses face. However, AI offers innovative solutions for these too.

AI-powered cybersecurity systems can safeguard data from breaches, while machine learning algorithms are capable of detecting potential threats early on, ensuring a robust defense at all times. Advanced training systems and simulations, powered by AI, can also help in upskilling the workforce, making them feel more comfortable and prepared for the rapid shift toward automation.

In essence, AI's role in enabling Industry 4.0 is indispensable, and its prevalence will only continue to grow. It's not merely about automation; it's about 'smart' automation where machines can learn, adapt, and improve. AI already lies at the heart of many current industrial processes and holds the keys to an exciting and dynamic future in manufacturing. So, as we stride into this future, understanding and harnessing its potential will be paramount for success in the new industrial revolution known as Industry 4.0.

Chapter 6. Case Studies: Successful AI Implementations in Manufacturing

Artificial Intelligence (AI) is no longer confined to the realms of science fiction, it is now a reality, profoundly impacting the manufacturing sector. Numerous industries have adopted AI technologies into their processes to reduce downtime, increase productivity, and improve their products' quality. Here, we delve into some of the most compelling case studies that illustrate the successful application of AI in Industry 4.0.

6.1. General Motors: Defect Detection with AI Vision

General Motors, one of the world's leading vehicle manufacturers, adopted AI-powered machine vision systems to inspect vehicles coming off the assembly line. The AI system, outfitted with an array of high-resolution cameras, can accurately and quickly detect any deviations in vehicles, such as misplaced components or minor cosmetic imperfections.

The inclusion of AI in their assembly process marked a significant improvement over their previous practice of manual inspection. It not only reduces the risk of human error, but the data gathered by the system can also be analyzed to identify recurring issues and implement preventive measures.

The implementation of this AI-powered system resulted in a significant decrease in defect rates while substantially improving assembly line efficiency. Their exercise is a classic example of technology augmenting human capabilities to deliver superior outcomes.

6.2. Siemens: Predictive Maintenance with AI

Siemens has been a vanguard in manufacturing innovation for over a century. In recent years, the global giant invested heavily in AI to modernize its operations — particularly in the area of predictive maintenance.

Their AI model uses sensory data to predict potential issues or malfunctions in their machinery before they occur. This preemptive approach enables the company to perform targeted maintenance, improving the equipment's lifespan and reducing unplanned downtime.

The system's AI model continuously learns from the accumulated data, honing its ability to identify anomalies and making its predictions increasingly precise. The success of Siemens' predictive maintenance solution has paved the way for widespread adoption of this technology, demonstrating the potential of AI to revolutionize maintenance procedures in the manufacturing industry.

6.3. Fanuc: AI for Automated Processes

Fanuc, a leading Japanese manufacturer specializing in automation, employed AI to enhance their production line's efficiency and adaptability. Their AI-powered system, FIELD (Fanuc Intelligent Edge Link and Drive), uses machine learning to analyze data collected from different devices and processes.

The advanced AI system generates insights to optimize production, such as identifying inefficiencies in the workflow and suggesting improvements. The system's adaptability allows it to cope with varying production requirements, with significantly reduced set-up times for new product variants.

The automated system's success catapulted Fanuc into global attention, demonstrating that AI's effective integration can enhance flexibility and efficiency in manufacturing.

6.4. Boeing: AI in Quality Assurance

Boeing, an aerospace titan, integrated AI into their quality assurance procedures via an image-recognition system. This innovative system is designed to spot defects or anomalies in production parts much faster than humans.

Furthermore, the system offers unparalleled precision and consistency—upholding Boeing's uncompromising commitment to safety. The integration of AI has led to a substantial reduction in inspection time, a simultaneous increase in defect detection, and has fundamentally enhanced Boeing's production quality.

6.5. Conclusion

The highlighted case studies provide a mere glimpse into the potential of AI in revolutionizing the manufacturing industry. These examples from some of the leading manufacturing corporations, illustrate the immense benefits of AI in various applications, from defect detection to quality assurance, predictive maintenance, and production optimization.

As AI continues to mature, such success stories and their resultant benefits will only multiply in the future. Manufacturers interested in remaining competitive should seriously consider the potential advantages of implementing AI in their own processes.

Chapter 7. Analyzing the Benefits: Efficiency, Accuracy, and Beyond

AI-based systems have, in recent years, become a fundamental pillar in the manufacturing industry, pushing efficiency and accuracy to previously unknown heights. This evolution has brought significant enhancements to what we might call traditional manufacturing processes, aligning the industry with the bold visions of the future that we often dreamed about, but hardly ever believed were attainable. The application of AI is that bridge between aspirations and reality - a leverage of technological advancement to the complexities of the production line.

7.1. Increased Efficiency and Reduced Wastage

Manufacturing industries have always been in a consistent pursuit to enhance efficiency at every level. With AI, this pursuit has accelerated at an unprecedented scale. AI-powered systems in manufacturing can rely on Machine Learning (ML) algorithms to parse enormous amounts of data and identify patterns much faster than a human would. Through this, they gain from timely preventive actions, predictive analytics, streamlined supply chains, and optimized energy use that significantly improves overall efficiency and reduces wastage.

One pivotal concept in this domain is predictive maintenance. This type of maintenance uses ML algorithms to predict equipment failure. By comparing present data to historical data, predictive maintenance enables systems to foresee equipment failures, thus reducing unplanned downtimes. Research indicates that applying this concept can reduce maintenance costs by up to 25%, halve downtime from equipment breakdown, and reduce repair time by almost 60%.

When AI-driven systems guide supply chain management, the efficiency reaches another level. Supply chains are often complex, involving multiple sequences - product planning, supply planning, demand planning, sales, and operations planning. AI's ability to recognize patterns and derive intelligent conclusions enables it to streamline these sequences, making the entire chain more efficient.

A noteworthy example would be the use of an AI system to monitor and control energy use in manufacturing processes. Not only could this optimise energy consumption by monitoring operation hours, load work, and unproductive time, but also significantly reduce the carbon footprint, leading to a more sustainable production process.

7.2. Greater Accuracy and Consistency

Accuracy is critical in manufacturing. Tiny mistakes can often lead to large losses. AI's ability to maintain consistency, reduce errors and continually learn from these errors can help drastically improve accuracy in manufacturing processes.

One locus of AI application aimed at increasing precision is in quality assurance. AI-powered robotic systems can play a significant role in recognizing and sorting defects in products within micro-seconds - a process that would take humans hours, or even days. Such tireless accuracy is simply unachievable through manual means.

What's more, these AI systems continually learn from their environment. By reviewing production-faults and rectifications periodically, they self-improve their process of prediction, making them less prone to overlooking or repeating mistakes - certainly an advantage in critical accuracy-emphasised processes.

In an industry, for example, where components must be meticulously assembled under strict quality assurance guidelines, leveraging smart robotic systems will guarantee an absolute level of consistency and precision. These systems continually cultivate their learning allowing for a gradual reduction in inaccuracies and progress towards almost perfection in tasks.

7.3. Autonomous Decision Making: Learning and Adapting

AI-enabled systems are not just about performing tasks faster or with more accuracy; it's also about autonomous decision-making. On top of performing routine tasks, these systems can also make decisions on-the-go by analysing wide-ranging datasets and taking an appropriate course of action.

Consider the management of raw materials in a factory. An AI-powered system, equipped with sensors, can gauge the levels of available raw material, appraise the rates of consumption, and make informed decisions about if, when, and how much to reorder.

An important aspect of autonomous decision-making is the ability to adapt to new situations. AI systems can analyze trends and patterns in historical data, predict future trajectories, and then adapt their course of action based on these predictions. This ability allows manufacturing units to respond quickly to market demands and changes in trends, capitalizing on the potential opportunities, or mitigating the severity of unexpected challenges.

AI's advantage of self-learning and continual adaptation reduces the turnaround time in decision-making processes, thereby providing an edge in the hyper-competitive market space of manufacturing.

These key areas -enhanced efficiency, increased accuracy, and autonomous decision-making -are bringing about revolutionary changes to the manufacturing industry's landscape. AI is not merely a tool, but effectively a production companion -always learning, forever improving, reducing waste, and ensuring consistency to the process. It's a journey towards ever-improving standards in manufacturing. AI marks the dawn of a new era in manufacturing -welcome to the Industry 4.0 revolution. With this knowledge, you too can be an active participant in this exciting revolution.

Chapter 8. Potential Pitfalls and Challenges in AI Integration

The integration of Artificial Intelligence (AI) into manufacturing—a key component of Industry 4.0—promises groundbreaking advancements, from the rise of smart factories to the introduction of self-optimizing production lines. However, this evolution is not without its challenges and obstacles. To equip businesses and stakeholders with the critical knowledge needed for this transition, we'll illuminate the potential pitfalls and how they can be addressed in a comprehensive manner.

8.1. Recognizing the Complexity of AI

The AI systems that are central to Industry 4.0 are of a complexity that is unparalleled in the history of manufacturing. With this complexity comes a steep learning curve and a multitude of potential complications and setbacks during integration.

AI systems are characterized by their ability to learn, adapt, and improve, distinguishing them from traditional automation technologies. For optimal performance, these systems require vast amounts of high-quality data, and proper 'training' of the system is critical. Inaccurate or incomplete data can lead to faulty predictions or misinterpretations by the AI, causing disruptions in manufacturing processes.

Addressing this challenge requires an investment in infrastructure and in-house expertise to ensure a well-rounded understanding of AI properties and mechanisms. Private or public courses and workshops can help to equip individuals with essential AI skills and competencies. Promoting the importance of data quality and proper system 'training' is critical to ensuring successful AI integration.

8.2. The Necessity of Cybersecurity Measures

Industry 4.0 rides on the back of digital technology, which brings AI systems alongside a need for robust and comprehensive cybersecurity measures. Connected machines and AI systems are vulnerable to malicious hackers, unauthorized personnel, and even unsuspecting employees who might accidentally breach protocols.

Designing and implementing resilient cybersecurity systems is fundamental for businesses. Not only do these systems protect sensitive company data, but they also ensure the safe and uninterrupted functioning of the AI systems integral to production lines. Regular cybersecurity audits, system patches, and extensive data encryption methods are some strategies to offset these cybersecurity challenges.

Investing in user education is equally essential, as even the best security measures are insufficient if employees are unaware of potential threats and appropriate response measures. Training programs on cybersecurity best practices can significantly bolster a company's overall defense mechanism.

8.3. Adapting to Changing Labor Market

AI and other Industry 4.0 technologies have already begun to reshape the labor market fundamentally, with certain roles becoming obsolete, and new ones emerging. With the rise of automated systems, job displacement has become a very real concern. However, while AI can replace certain tasks, it also paves the way for new roles that require human intelligences such as creativity, strategic thinking, and emotional intelligence.

Addressing this challenge involves preparing the workforce for the 'future of work.' On the one hand, there is the necessity of providing training for employees to adapt to new roles and tasks created by AI integration. On the other hand, talent recruitment and retention strategies must adapt to attract employees with the skills required by Industry 4.0. This underscores the need for strong partnership between corporations, educational institutions, and governments to create pathways for skill development and lifelong learning opportunities.

8.4. Ensuring Regulatory Compliance

Industry 4.0, with its blend of physical and cyber systems, poses unique regulatory and compliance challenges. There are questions about data privacy, safety standards, intellectual property rights, and international trade complexities—all requiring legal and ethical guidelines that can adequately reflect the new realities of manufacturing.

To engage with these issues, industries should actively contribute to the dialogues and debates about AI regulation. Collaboration between companies, policymakers, and academics is vital to developing regulations that take into account the intricacies of AI systems while ensuring fair competition and innovation in Industry 4.0.

8.5. Overcoming Resistance to Change

Finally, resistance to change can be a significant roadblock in the journey towards AI-integrated manufacturing. Lack of understanding, fear of job displacement, or mistrust towards machines can lead to resistance among employees, making the transformational process more challenging than it should be.

Companies can counteract these sentiments by fostering an inclusive environment where the shift towards an AI-integrated system is collectively carried. This includes regular communication about the benefits and opportunities presented by AI, reinforcing a clear vision for the future, and outlining each employee's critical role in this new frontier.

In conclusion, the integration of AI within manufacturing is undoubtedly a complex task, colored with potential pitfalls and challenges. However, with in-depth understanding, careful planning, and proactive strategies, these issues can be adeptly navigated. The reward—a revolutionized manufacturing industry—is potent with opportunities for exponential growth, efficiency, and innovation.

Chapter 9. Sustainable Manufacturing: The Role of AI in Green Industry Practices

As we propel ourselves towards a more progressive era of advanced machinery and intelligent systems, one of the major aspects garnering significant attention is sustainable manufacturing. In this chapter, we delve into how Artificial Intelligence (AI) is pivotal in facilitating green industry practices, steering us towards a more eco-friendly and sustainable future.

9.1. The Intersection of AI and Sustainable Manufacturing

Sustainability and AI may seem diametrically opposed concepts, but they collide and integrate uniquely in the modern manufacturing landscape. AI allows us to generate data-driven insights that can be applied to optimize operational efficiency, minimize waste, and reduce the environmental impact of industrial systems.

AI technology offers advanced analytics capabilities that can be used to discern patterns and correlations that would otherwise go unnoticed. These insights can be applied to refine manufacturing processes, resulting in fewer resources consumed and less waste produced. For example, AI can predict machine wear and tear, allowing for proactive maintenance, minimizing unscheduled downtime, and reducing the waste of resources.

9.2. AI Applications in Energy Efficiency and Waste Reduction

AI-powered systems play a crucial role in energy efficiency and waste reduction, two decisive factors in sustainable manufacturing. Energy consumption is one of the most significant costs for most manufacturers. AI can help optimize energy use through predictive analytics, calibration of machinery, and intelligent control systems.

For instance, AI-driven energy management systems can continually monitor and analyze a factory's energy consumption patterns. By applying machine learning algorithms, these systems can identify inefficiencies and adjust electricity usage on the fly, reducing overall energy consumption.

On the waste reduction front, AI is equally powerful. AI-powered systems can track, categorize, and manage waste - from its creation to its disposal. The intelligent classification of waste aids in the segregation and recycling process, while predictive models can forecast waste generation, providing manufacturers with insights to optimize waste management practices.

9.3. AI in Resource Conservation and Recycling

Industrious application of AI can create a closed-loop system in manufacturing, where waste is minimized, and resources are recycled or reused as much as possible. With AI, raw materials can be used more efficiently, machines can be operated optimally, and even waste can be repurposed effectively.

For example, AI techniques like machine learning can help manufacturers better predict their resource needs. More accurate predictions mean fewer wasted resources on overproduction or excess inventory. When it comes to recycling, AI can automate the segregation process, sorting recyclable materials more accurately and swiftly than human-led operations.

9.4. AI and Renewable Energy in Manufacturing

The use of AI extends to the application of renewable energy resources in manufacturing. AI algorithms can help optimize the use of renewable energy by predicting energy production based on environmental factors. For instance, AI can forecast solar power generation based on weather patterns and optimize the consumption of generated energy.

AI-powered micro-grid systems can manage and balance energy resources to ensure steady power supply during peak load times and minimize grid instability. This autonomous control not only increases the reliability of renewable energy sources but also saves costs for the manufacturer while significantly reducing carbon emissions.

9.5. Embracing AI for a Sustainable Future

AI bears the promise of a green and sustainable future for manufacturing. As our understanding of AI evolves, so too does our ability to harness it to create more sustainable practices in the manufacturing industry. AI tools like predictive analytics, machine learning, and intelligent control systems have the ability to revolutionize how resources are utilized efficiently, and waste is effectively managed.

By embracing AI, manufacturers can not only improve their bottom line but also significantly minimize the environment footprint of their operations. This not only benefits the planet but aligns with evolving consumer sentiments, where sustainability and social consciousness are increasingly considered alongside price and quality.

In conclusion, AI is set to play a pivotal role in the sustainable manufacturing practices of the future. Its countless applications, ranging from energy efficiency, waste reduction, resource conservation, recycling, and renewable energy, have the potential to bring about a paradigm shift in the industry. This confluence of AI and sustainability promises not just a sustainable future for manufacturing, but a green and bright future for our planet.

Chapter 10. Preparing for the Future: Upskilling the Workforce for AI-Integrated Manufacturing

As we stand on the precipice of an AI-driven industrial revolution—tagged Industry 4.0—the skills required by employees are changing. Manufacturing firms must respond effectively to this transformation by investing in upskilling and training programs that equip their workforce for a future intertwined with AI.

10.1. The Imperative for Upskilling

In the face of rapid technological advances and amidst the increasingly prominent role of AI, companies need to seriously consider the skills their workforce will require in the near future. Industry 4.0 demands "smart factories" that integrate automation, data exchanges, AI, and manufacturing technologies. This shift necessitates an equally smart workforce adept in managing these advanced technologies.

At the outset, it's essential to understand that technology adoption shouldn't overshadow or invalidate human capacity. Instead, the spirit of upskilling in the context of Industry 4.0 is based on harmony—one where AI and humans coexist and complement each other's skills.

Studies predict that by 2022, the core skills required for most jobs will change by 42%, according to the World Economic Forum, with complex problem-solving, critical thinking, and creativity at the top. This landscape presents an interesting challenge: How do we prepare our workforce for this imminent future?

10.2. Understanding the Skills Gap

Before launching into upskilling initiatives, companies should identify and understand the "skills gap." The term refers to the difference between the skills companies need for efficient operation and successful growth, and those that their employees currently possess.

Critically, this gap can't be comprehended as a one-time analysis. As technology evolves, so does the nature and breadth of the skills gap. Companies must regularly audit and assess their employees' abilities, aligning them with current and future industry requirements. From these audits, organizations can determine the specific training and upskilling measures they need to implement.

10.3. Plugging the Gap: Essential AI Skills for Workers

With the identified skills gap in hand, companies can begin to develop targeted programs that address the required competencies for an AI-integrated future. Here, we briefly discuss the key skills that your workforce needs to acquire.

1. Technical skills: These include skills in programming languages used for AI and machine learning applications like Python, R, and Java. Workers should also understand data modeling, evaluation techniques, and system design.

2. Data literacy: Given the centrality of big data for any AI implementation, employees need a working knowledge of data analysis and interpretation, data cleaning, and data security protocols.

3. Soft skills: Despite the technical nature of AI, communication, problem-solving, collaboration, and emotional intelligence will become even more critical. As AI takes over routine tasks,

employees who can manage complex problems, lead teams, and communicate well will be increasingly valuable.

4. AI Ethics: As AI becomes more integrated into daily operations, ethical considerations like transparency and fairness, privacy, and accountability will be of central importance.

10.4. Strategies for Upskilling

Having addressed the crucial skills needed in the age of AI, we now turn our focus on strategies firms can adopt to achieve upskilling.

1. On-the-job training: It allows employees to earn while they learn, reducing downtime and enhancing productivity. These training sessions can be tailored to the company's specific needs and delivered at an appropriate pace.

2. Collaborate with educational institutions and online platforms: Partnerships can be formed with universities, colleges, Nanodegree providers or Massive Open Online Courses (MOOCs), offering specialized courses in AI and related areas.

3. Mentoring and Peer-to-peer learning: Encourage experienced employees to act as mentors, fostering a rich culture of learning within your organization.

4. Sponsor certifications and further education: This strategy can help companies retain talented employees while ensuring they acquire the skills necessary for the future.

Adopting these strategies demands investment—of time, money, and leadership focus—but the potential returns of crafting a skilled workforce ready for Industry 4.0 are astronomically high.

10.5. Embracing a Culture of Lifelong Learning

Effective upskilling is not just a series of training programs. Instead, it demands nurturing a culture where learning is appreciated, rewarded, and considered an ongoing process. Emphasize the importance of constant learning and embed it in the organization's culture and values.

In conclusion, the wave of Industry 4.0 and AI isn't coming—it's here. Preparing for this future means equipping your workforce with the right skills, fostering a learning culture, and bridging the skills gap effectively. By doing so, companies can ensure they remain competitive, innovative, and ready to leverage the vast potential that AI-powered manufacturing offers.

Chapter 11. Conclusions and Future Trends: The Horizon of Industrial AI

In the energetic pursuit of Industry 4.0, manufacturing companies worldwide have made Artificial Intelligence an inherent part of their processes. This adoption has fostered new efficiencies, enhanced production quality, and cultivated impressive industrial results. As we look beyond the present status quo towards the horizon of industrial AI, it's clear that we're on the cusp of an even more transformative era. However, to derive a comprehensive understanding, we need to start by deciphering both the pervading and upcoming trends in this dynamic landscape.

11.1. The Maturation of AI in Manufacturing

We've made leaps beyond merely installing AI algorithms within our production lines. Organisations have instead started to understand AI as an omnipresent tool, central to every operation. Advanced machinery and AI-driven robotics working in concert have already demonstrated their potential in automating redundant tasks and precisely executing complex ones. The growing maturation of AI has also begun to reflect in its more subtle applications like predictive maintenance, resource optimization, and just-in-time manufacturing.

It's a fascinating trend to watch as firms move away from viewing AI as a separate, 'added-on' feature. Instead, it's gradually evolving into a widespread necessity, streamlining operations and bolstering efficiency. The future of the manufacturing industry undoubtedly lies in this deeper integration of AI throughout all segments of the production chain.

11.2. Decentralization and Edge Computing

Just as we see a trend towards AI integration, there's a clear shift towards decentralization. The application of edge computing in manufacturing enables data processing closer to the source, minimizing latency and enhancing real-time decision-making. Over time, the decentralized processing of data will enable more autonomy for AI systems, further enhancing their efficiency and responsiveness.

In a highly distributed manufacturing landscape, this evolution will enable the realization of responsive and intelligent 'factories of the future'. These facilities will be interconnected, IoT-enabled, and AI-optimized, providing a semblance of autonomy while making operations more efficient.

11.3. Flexible Automation and Customization

One of the significant strides in manufacturing involves overcoming the 'one-size-fits-all' approach—thanks to AI and automation. Rather than mass-producing standardized products, factories can now use flexible automated systems to cater to custom orders and smaller product runs. The ability to adapt quickly to market demands and shifts is an enormous stride forward for the sector.

In the future, we won't just see tailored products; we'll also see flexible manufacturing lines. Here, reconfigurable robotics and advanced AI algorithms will work together to swap from manufacturing one product type to another swiftly.

11.4. AI-Driven Quality Assurance

Manufacturers across the globe grapple with the perennial issue of quality control. Knockout blow of AI lies in its capabilities of forecasting defects prior to their occurrence, reducing wastage and rework significantly.

Advances in machine vision technology, a subset of AI, are set to revolutionize quality checks on production lines. Using high-definition cameras and sophisticated image recognition algorithms, manufacturers can detect even the smallest inconsistencies or faults in components or products. Looking forward, the ubiquity of AI will redefine quality assurance standards around the world.

11.5. Labor and AI Synergy

In the discourse about AI in manufacturing, one cannot ignore the human factor. Despite predictions of job losses due to automation, the goal is not to replace human labor, but rather, it's to create a synergy between labor and AI.

AI will predominantly take over tasks that are hazardous or gratingly repetitive—things humans shouldn't do. Meanwhile, human workers will transition to overseeing the AI systems, equip themselves with new age skills, and take up roles that need advanced decision-making, creativity, or the 'human touch'. The forthcoming synergy between AI and human labor will revolutionize manufacturing, leveraging the strengths of both.

11.6. Ethical Implications

As we incorporate AI deeper into our systems, corresponding accountability and ethical considerations also surge. Moving ahead, the need for transparent AI decision-making pathways and ensuring ethical handling of data collected from AI systems will dominate conversations around technology implementation. Businesses pioneering in AI must consider the privacy, trust, and morale implications of AI applications while planning future strategies.

In conclusion, our journey through the current AI implications and future trends in manufacturing unravels a game-changing narrative. As the industry moves forward, the importance of AI is not just confined to 'robots'; it plays a vital role in every facet of the manufacturing process. Only by fully embracing the transformative power of Industry 4.0 can manufacturing businesses touch the zenith of striving efficiency, quality, and productivity targets. As we step into the impending era, every manufacturing professional needs to become an active participant and informed decision-maker. After all, the future belongs to those who create it, and the AI-led manufacturing revolution is ours to build.

www.ingramcontent.com/pod-product-compliance
Lightning Source LLC
Chambersburg PA
CBHW071033260726
48661CB00007B/3015